Solar Photovoltaic System Economic Analysis for the Electric Energy Consumer

DAN CATLIN

Copyright © 2013 Dan Catlin

All rights reserved.

ISBN-13: 978-1483995441

ISBN-10: 1483995445

SOUTHWEST PUBLICATIONS 2013

DEDICATION

This thesis is dedicated to the principal that we can use energy to improve our lifestyle without damage to the environment. I am grateful for government programs and free enterprise competition that has reduced the cost of solar photovoltaic (PV) systems and energy storage systems. In many locations, solar PV systems are now cost competitive with energy from utility scale central power plants. As electric rates continue to rise and the cost of solar PV systems continue to drop, I expect major changes in where we obtain electricity. The future is bright!

CONTENTS

ACKNOWLEDGMENTS   i

1. **Chapter One: Introduction**   1
2. **Chapter Two: Energy Production**   3
3. **Chapter Three: Return on Investment (ROI)**   10
4. **Chapter Four: Tax credits and other incentives**   15
5. **Chapter Five: Other things to consider in your analysis**   17
6. **Chapter Six: Conclusion and where to find help**   21

ACKNOWLEDGMENTS

I am grateful to my wife, Carol Catlin, for her encouragement, editing skills, and support on this project. Mitch Waite has helped with formatting the document. SRP, our electric utility; NREL, National Renewable Energy Lab.; and others have provided information and support.

CHAPTER 1: INTRODUCTION

Although a solar energy enthusiast, I recognize that using solar energy for your home or business does not necessary make good economic sense. I will explain the factors that should be evaluated to determine in specific situations if installing solar PV (Photovoltaic) is a wise economic decision. Be aware that the recent advances in solar PV systems and cost reductions should be taken into account. One should avoid generalizations about the economic advantages or disadvantages of installing solar PV.

Factors to consider:

1. How much energy will the system produce? Details about how to figure this out are in chapter 2. Hint: your location is very important…how much sunshine does your area receive, and what shading issues are involved.

2. What Return In Investment (RII) is expected? The case being the cost avoided in your electric utility bill is a RII. In most situations, this will be an investment in which the RII is not taxable. What is the expected life of the proposed system and rate of degradation? What about insurance and warrantees? What is the cost of money assuming one borrows part or all of the capital to install the system? One needs to consider the

interest rate of borrowed capital, the loan time frame, and payment schedule. I discuss these issues in chapter 3.

3. What tax credits, tax incentives, and utility incentives are available in your area? How do tax credits and utility incentives influence the choice between leasing a solar PV system; entering into a Purchase Power Agreement (PPA) for obtaining electric energy from a solar PV system; or purchasing a PV system? Discussed in Chapter 4.

4. Do you want to consider energy cost inflation in your analysis? What about utility power outages or off grid systems? Have you looked at your utility rate schedule specifics? Do you want to include the benefits of additional points for Leadership in Energy & Environmental Design (LEED) certification? Would you want to install pole-mounted systems to provide shade for parking or other purposes? Do you want to do part or all of the installation yourself? Where do you find qualified solar contractors? These topics will be discussed in chapter 5.

5. Conclusion. Do you need help figuring it out? Who can you trust? Where to look for help? Discussion in chapter 6.

As an environmental specialist at Fort McDowell Yavapai Community, I have specified and overseen the installation of several solar PV systems. Having lived in Arizona for over thirty years and over five years using solar energy at our home, I wanted to share our experiences with using solar energy.

ROOF MOUNTED SOUTH FACING SOLAR MODULES ON A SMALL OFFICE BUILDING NEAR PHOENIX, ARIZONA

CHAPTER 2: ENERGY PRODUCTION

Electrical energy is usually measured in units of watt-hours or kilowatt-hours. A kilowatt hour is 1000 watt hours and will be noted as KWhr. If you operate an appliance that draws 500 watts for 4 hours that appliance used 500 X 4 = 2000 watt hours = 2KWhr. Some electric rate schedules also have charges based on peak power demand used as well as total energy used. Peak demand would be the highest watt usage recorded during a specified time period. Solar

PV systems can significantly reduce the amount of energy you need to purchase from your electric utility but usually do not significantly reduce your peak demand. For example, if your peak demand occurs at night, your solar PV system will not reduce your peak demand. If the peak demand usually occurs during the middle of the day, then after installing a solar PV system, the peak demand will likely occur during a different time period (when the sun is not shining).

How much energy will a solar PV system produce? Hint: your location is very important…how much sunshine does your area receive and what shading issues are involved. There are websites that have calculators that will help you answer production questions and estimated system costs as well as estimated value of energy produced. I like to use the National Renewable Energy Lab (NREL) websites:

http://rredc.nrel.gov/solar/calculators/PVWATTS/version1/

http://maps.nrel.gov/imby

The first website allows you to click on your state and than a city near you. Then you need to enter some information about the system you are considering: KW rating; DC to AC derate factor (I suggest using .95 or higher. The .77 value is way too low from my experience); array type (a pull down list); array tilt; orientation in regard to south (180^0 would be due south and is the default value); and your cost of electricity in cents. For most energy production on a year around basis the tilt of the panels would be the default (your latitude). More tilt in the winter would give higher production in the winter. Less tilt in the summer would give more production during those months. Some pole-mounted systems can have the tilt adjusted by the season. I recommend a tilt of at least 10 degrees so that rains will tend to drain and clean the panels. Normally the panels should not need cleaning (any cleaning cost would offset any additional energy production credit).

There are a number of existing solar PV systems that have been producing electricity for many years without maintenance or repairs and with minimal degradation (lower production with age). In fact, panel manufacturers normally warrantee their products for 25 years regarding energy production.

Sun tracking solar Photovoltaic system powering an air quality and metrological monitoring station

DAN CATLIN

Example production history for two years vs. projected, according to PV watts calculator:

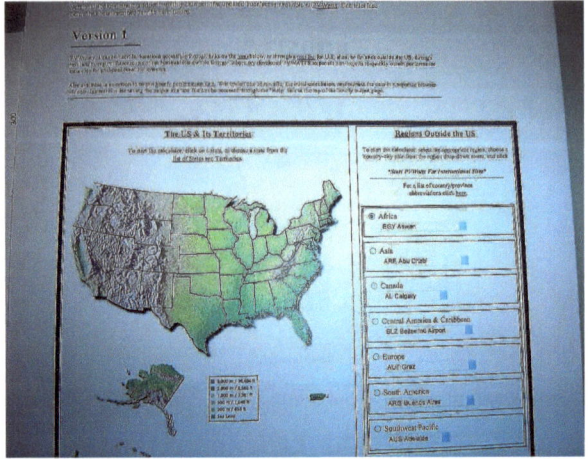

http://rredc.nrel.gov/solar/calculators/PVWATTS/version1/

SOLAR PHOTOVOLTAIC SYSTEM ECONOMIC ANALYSIS FOR THE ELECTRIC ENERGY CONSUMER

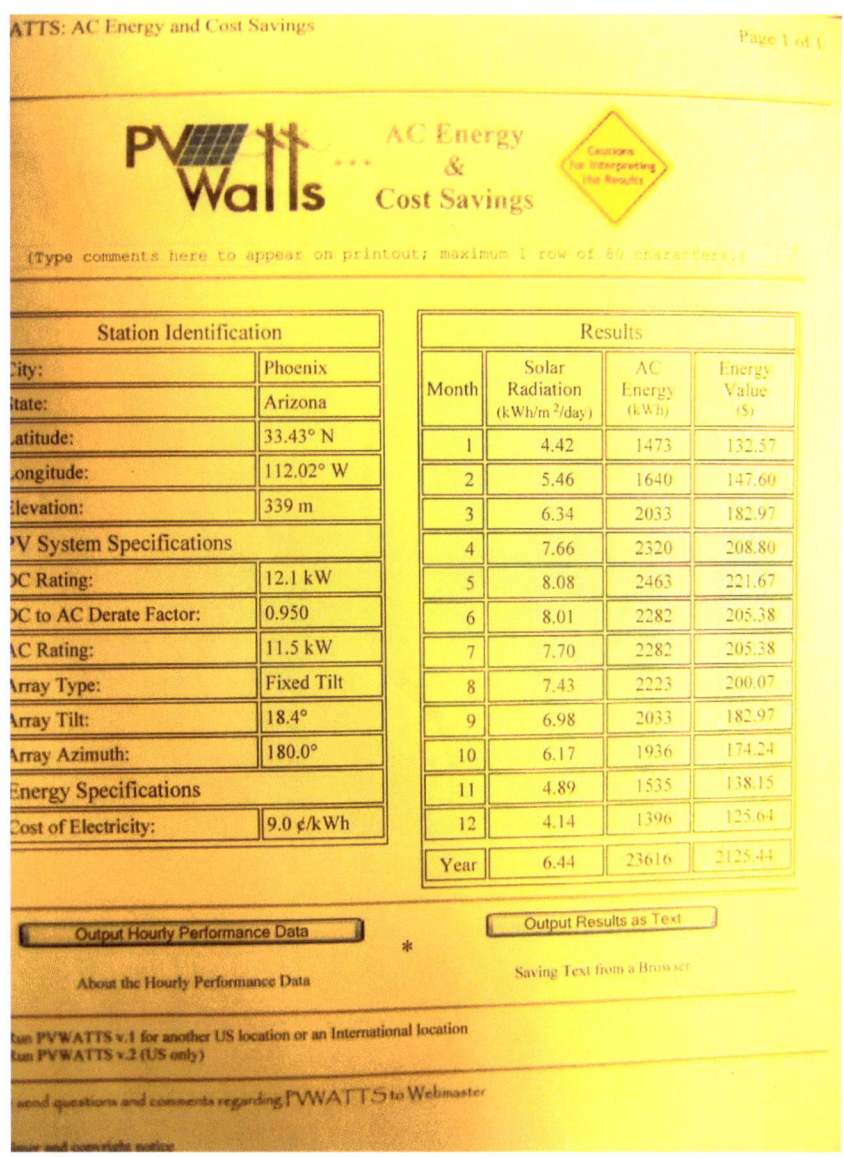

Small office building
Near Phoenix, Arizona

PV Watts projected 12.1 kw system, 18.4° tilt, array azimuth 180°

	KWhr according to utility production meter	KWhr according to Enlighten production meter	AC energy kwh AC/DC derate: 0.95	12.1 kw system 18.4° tilt array azimuth 180° AC energy kwh AC/DC default Derate: 0.77
year 2011				
January	1679	1730	1473	1190
February	1587	1630	1640	1327
March	2082	2150	2033	1646
April	2196	2280	2320	1883
May	2392	2480	2463	1998
June	2238	2330	2282	1850
July	2133	2220	2282	1849
August	2081	2170	2223	1802
September	1903	1980	2033	1646
October	1849	1910	1936	1568
November	1455	1500	1535	1239
December	1335	1370	1396	1126
2011 totals	22,930	23750	23616	19124
year 2012				
January	1511	1560	1473	1190
February	1707	1770	1640	1327
March	2121	2200	2033	1646
April	2146	2230	2320	1883
May	2373	2480	2463	1998
June	2240	2340	2282	1850
July	1966	2060	2282	1849
August	1981	2080	2223	1802
September	1827	1920	2033	1646
October	1922	2010	1936	1568
November	1591	1660	1535	1239
December	1407	1470	1396	1126
	22,792	23780	23616	19124

The above table shows the first two full calendar years of energy production form a 12.1 KW system installed March of 2010. The solar PV system in this example is on the small office building shown on page 2. This system produces over 20% of the energy used by this building. Several things to note regarding this example:

- Less than 1% difference in energy production between year 2011 & 2012 according to the production meter installed by SRP Electric utility.

- The enlighten monitor actually shows more production the second year.

- The production meter was located next to the billing meter and was about 200 feet from the building (there could be some line loss)

- The enlighten meter was located in the building and measured the energy produced by each solar panel.

- The assumption using 0.95 for the DC to AC derate factor rather than the default factor of 0.77 gives a much closer approximation to the actual production history.

I have observed that cloud cover can significantly reduce energy production. Heavy cloud cover over most of the day can reduce that day's energy production by 75% or more. Fortunately in Arizona, we mostly have clear sunny days. When we do have clouds they seldom stay around for more than part of a day. I am confident that weather variations by month and between years accounts for most of the energy production differences in the table above.

The solar industry recognizes that there is some energy loss as ambient temperatures rise. This is due to voltage drop with higher temperatures and is often documented in published panel specifications. In other words colder temperatures with the same amount of sunshine results in higher production. Even with the hot summers we have in the Phoenix, Arizona area, we can still get significant energy production during our clear sky's during the summer. You can expect more production at higher elevations with cooler days such as we have in Flagstaff, Arizona.

CHAPTER 3: RETURN ON INVESTMENT (ROI)

There are a number of reasons, including emotional reasons, for installing solar PV to provide a portion or all of your electrical energy usage at your home or business. There are economic reasons why a number of major corporations are installing solar PV systems at some of their locations (Wal-Mart, Google, Ikea, and many others). Does installing solar PV make economic sense for you? If your investment of $10,000.00 results in $1,000.00/year in savings on your electric bill (10% ROI), would that be adequate for you? How would this ROI compare to other capital investments you consider? When comparing expected ROI from various options what criteria do you use? These are questions you need to consider.

When interest rates are low, and you do not need to use much of your own money up front, the potential for much greater ROI exists for your investment. For example if you can finance a $100,000 project with 10% down and the balance at 3% interest over a 20 year period:

http://www.amortizationtable.org/

Assume that investment of $100,000 would pay for a 25 KW solar PV system ($4/watt installed) in the Phoenix, Arizona area. Using PV watts calculator (see chapter 2) for Phoenix, Arizona and assuming a value of $0.15/KWH over a 20 year period, south facing

SOLAR PHOTOVOLTAIC SYSTEM ECONOMIC ANALYSIS FOR THE ELECTRIC ENERGY CONSUMER

array, 10 degree tilt, and 0.95 DC to AC derate factor the value of the energy produced would be $140,931.00. The principal & interest over 20 years would be $119,793.08. Under these conditions, the investment of $10,000 would give a tax free return of $21,137.92 - $10,000.00 = $11,137.92 over the 20 year period as well paying off the original loan. That would give an average 11% per year return on your investment. When combined with possible tax credits and utility incentives, the ROI is even better.

Lets try another example using different assumptions:

1. 5000 watt system (5 KW) for residential home.
2. Location is Phoenix, Arizona
3. $5 per installed watt cost ($25,000.00)
4. 30% federal tax credit
5. $1,000.00 State income tax credit
6. $0.50 per installed watt utility rebate
7. Value of electricity is $0.12 per KW-hr.
8. South facing panels at 10 degree tilt

$25,000 - $2500 utility rebate = $22,500.

$22,500 - $6,700 (30% federal tax credit) = $15,750

$15,750 - $1,000 state tax credit = $14,750 home owners cost

Value of energy produced per year = $1127 (see chapter 2 for how to calculate expected value of energy produced).

Home owners ROI = 7.6%

Since that is based on savings that would be tax free! Over the life of the system (25 years +), I bet the average cost of electricity will be much higher than 12 cents per KW-hr. That would lead to an even better average ROI over the years. Now, if the home owner has a home equity line of credit (HELOC) with an interest rate of 6% in which to finance the project, he could finance the project with no up front money, claim the tax credits, obtain the utility rebate, and lower his electric bill. If he itemizes his deductions on his federal and state tax returns, he can also deduct the interest expenses due to his HELOC loan.

When looking at electric costs and electric rates, you need to use the energy cost in cents/kilo watt hour (cost per KWhr) when estimating how much solar energy systems will save. Many rate schedules include a connection charge and other charges not based on energy usage. For example some rates include a demand charge based on the peak KW demand during the billing cycle. Depending on when the peak demand (usage) occurs, solar PV systems are not likely to significantly reduce the demand charge. For detailed analysis, you also need to look at time of day usage rates as well as different rates depending on the season of year and your particular rate schedule.

Business owners can also offset the installation cost of solar PV systems with tax credits as above and also reduce their tax liability with depreciation (discuss this with your tax advisor/specialist) if the system is used for their business.

As of this writing (February 2013) present law regarding the 30% federal tax credit expires December 31, 2016. Utility rebates have been reduced a lot since we installed solar PV at our home. Most of what we had installed, the utility paid a rebate up front of $3.00/watt. On the other hand, installation cost for systems have been reduced substantiality. Most of the reduction has been due to the drastic reduction in the cost of the solar PV modules. Competition among installers has also helped lower cost.

SOLAR PHOTOVOLTAIC SYSTEM ECONOMIC ANALYSIS FOR THE ELECTRIC ENERGY CONSUMER

To obtain costs of equipment and extensive information about solar modules; inverters (which convert the DC voltage from the modules to AC voltage); racking equipment; monitoring equipment; wiring; breakers; and voltage regulators there are a couple of companies I refer to:

Wholesale Solar: www.wholesalesolar.com

http://www.wholesalesolar.com/why-wholesale-solar.html

(800) 472-1142, or email them at sales@wholesalesolar.com

Solar Penny: www.pennysolar.com

(480) 374-3230, or email them at Sales@SolarPenny.com

These websites are very easy to use and have contact phone numbers & email addresses from which you can receive additional help.

I have also found some of the solar installation contracting companies very helpful with information.

Some of the contractors I have dealt with here in Arizona or received quotes from include:

- Dependable Solar:
  http://www.dependablesolarproducts.com/

  Lane Garrett, Tempe phone (480) 967-7781 or toll free (866) 967-7781

- S3 Energy: http://s3energy.com/

  Todd Smith, phone (480) 399-1476

- Royal Covers: http://www.royalcovers.com/

  Tanner Bishop, phone (480) 926-2300

- Solar City: http://www.solarcity.com/

    Phone (888) 765-2489

- SPG Solar: http://www.spgsolar.com/
- Wilson Electric: http://www.wilsonelectric.net/
- American Solar Electric: http://www.americanpv.com/

And there are also many more licensed solar contractors working in Arizona. Some of the above contractors also work in other states as well.

CHAPTER 4: TAX CREDITS AND OTHER INCENTIVES

What tax credits, tax incentives, and utility incentives are available in your area? How do tax credits and utility incentives influence the choice between leasing a solar PV system; entering into a purchase power agreement (PPA) for obtaining electric energy from a solar PV system; or purchasing a PV system?

Here again one can learn much from the Internet. You just need to know where to look. Some of my favorite websites include:

http://www.dsireusa.org/

http://www.arizonagoessolar.org/

The 30% federal tax credit is a big incentive for those who have federal tax liability. If you are not familiar with how to take this credit or the depreciation allowances you should consult with a tax consultant. You can also obtain information from the IRS through the website: http://www.irs.gov/

In general, if you are not able to use the tax credits and depreciation allowances, you would probably be better off going into a lease or

PPA agreement with an investor that can use those credits and allowances. A lease would probably give you a better deal, especially if the lease included guaranteed energy production with lease payment reductions based on reduced production. A purchase power agreement (PPA) gives you confidence about what you will pay for the electricity produced by the solar PV system. The owner has an added incentive to do any maintenance needed to keep energy production high. One common lease or PPA agreement includes a purchase option after a specified number of years. This gives the property owner the ability to purchase the system at a reduced cost after the original investor(s) are able to use up the tax credits and depreciation allowances.

CHAPTER 5: OTHER THINGS TO CONSIDER IN YOUR ANALYSIS

Do you want to consider energy cost inflation in your analysis? What about utility power outages or off grid systems? Have you looked at your utility rate schedule specifics? Do you want to include the benefits of additional points for LEED (Leadership in Energy & Environmental Design) certification? Would you want to install pole-mounted systems to provide shade for parking or other purposes? Do you want to do part or all of the installation yourself? Where do you find qualified solar contractors?

When doing the financial analysis, you should be familiar with your electric utility rate schedule. Some rate schedules base the bill not only on the KW-hours of energy used but also on the peak KW demand as well as the time of day energy usage. Solar PV will reduce the KW-hours of energy usage but may not impact significantly the demand charge.

It is not possible to predict what the electric rates will be in the future but the consensus is that rates will increase. During economic hard times, utilities try extra hard to keep rates from increasing. They tend to postpone preventive maintenance and delay capital investment in new production and transmission systems. Eventually these issues have to be addressed. Many of the existing power generation stations are 25 years old or older. Environmental regulations require best available air pollution control equipment to be installed on new fuel burning power plants or plant upgrades. Inflation in labor and equipment costs also will be reflected in the electric rate inflation. Fuel costs tend to increase over time. For utilities to stay in business, these and other costs need to be accounted for when utilities request rate increases. Utilities monitor electric demand and many other system conditions and try to predict future needs by looking at system history as well as economic forecasts. Planning and preparing for expected growth in electricity usage from the utility service area is very important to the utility as well as the community. Because most of society expects (or needs) electricity to be available at all times, excess generation capacity and redundant systems are built. These add to the cost of operating an electric utility so that when demand peaks or equipment fails, the public still receives electricity. As the economy recovers from recessions the demand for electricity increases.

For off grid considerations, you need to consider the cost of bringing electricity to your property versus generating your own electricity. Off grid solar PV systems need some type of energy storage system (usually batteries) and this adds to the cost and maintenance. It also influences the choice of inverter type and also voltage regulation required. When considering the cost of electricity for off grid areas when line extension to a nearby electric utility is too expensive or not available several options need to be considered. Then you need to compare the cost of electricity from a solar PV system with storage to the cost of using a fuel burning generator with fuel storage; wind powered system, with energy storage; possibly a small hydro powered generator, with some energy storage system; or some combination of the above.

SOLAR PHOTOVOLTAIC SYSTEM ECONOMIC ANALYSIS FOR THE ELECTRIC ENERGY CONSUMER

For grid tied systems you also have the option of energy storage systems and/or emergency generation systems for when you lose power from the utility. Most grid tied systems do not produce energy when you lose power from the utility. Now, the question is how much of your energy demand should the storage and/or emergency generators be designed to provide and for how long of a time period? Does your utility have frequent and prolonged black outs or brown outs? What are your critical electric needs? Do you want to take advantage of lower utility rates in exchange for having your being the first to have your electricity cut when the utility has problems meeting their customer's demands? Backup systems are expensive but may be worth it depending on your needs and the utility reliability.

An apartment complex in Mesa, Arizona provides shade for their tenants vehicles using photovoltaic panels. An additional incentive for people to rent here is that electricity usage is included in their set monthly rental payment.

No doubt, the apartment management were able to use production based utility rebates, federal and state tax credits and depreciation allowances to help provide this shade and electricity to their tenants

CHAPTER 6: CONCLUSION AND WHERE TO FIND HELP

Solar PV systems installed by electric utility consumers will result in less energy needed to be purchased. Usually experts recommend that energy conservation measures be undertaken before installing solar energy equipment. Some energy conservation measures can be incorporated with low or no expense. For example, you could raise the thermostat during hot weather and lower the thermostat during cold weather. By appropriately timing the use of outside air for ventilation you can reduce the use of HVAC systems. Most fans are of low cost and use very little electricity. Adequate air movement can reduce the amount of cooling needed. For dry climates, evaporative coolers are a low energy way to cool. People have different temperature expectations. When some people are using heaters to warm themselves while the buildings air-conditioning system is trying to cool the building, a lot of energy can be wasted. Added insulation, radiant barriers, and energy efficient windows & doors when installed properly will save energy. Old refrigerators, freezers, and heat pumps are often energy wasters. Replacement with energy efficient units may be cost effective. Replacement of high wattage lights with lower wattage lights with the same lumen output is also a good decision in most cases.

Some utilities provide financial aid to consumers who replace old appliances and HVAC systems with new energy efficient equipment.

Do you need help figuring it out? Who can you trust? Where to look for help? I always recommend that people decide what they want to do before talking with a salesman. A lot of the solar salesmen will want you to sign up with them before they leave your home.

You know the pitch: "this deal is only good today". If you know the market, know your situation, and what you want, you will be better prepared to meet with these salesmen and maybe sign the contract before they leave. But, if you have difficulty analyzing your situation or do not want to take the time to figure it out, you may decide to consult with someone more familiar with your utility and contractors in your area. The person you retain should not also be a salesman for the company you decide to purchase a solar PV system from. Your electric utility company may be able to help. I live in Mesa, Arizona and our electric utility, SRP, has been very helpful with information about solar energy systems. They have made available information about average installation costs of solar systems installed locally. They have also provided information about various types of solar energy systems. SRP has also provided financial incentives for installing solar energy systems by their customers either as an up front cash rebate or a production based rebate. These utility financial incentives have been reduced substantially since 2010 and are likely to be eliminated in the future. Due to the political climate encouraging utilization of renewable energy systems, it is likely that your local electric utility will also be helpful.

Electricity is very important for modern lifestyle. Where do we save without impacting business or home lifestyle negatively? Is it reasonable to produce our own electricity? I set out to try this in Mesa, Arizona over 5 years ago. We live in a very hot climate and need a lot of electricity to run our heat pump day & night for a number of months every year. We also operate a hot tub and exercise pool year around at our all electric home. It has been over 3 years since we have purchased electricity from SRP. Our excess electric energy production pays the utility connection fee for about half of the year (they only credit us at the wholesale rate for our excess generation).

ABOUT THE AUTHOR

Dan has 30 years experience in the environmental field and has followed energy issues for most of those years. From 1980 to 1982 he did energy use and cost studies at the world's largest underground copper mine near Mammoth, Arizona operated by Magma Copper Company. He has watched the transition from use of fuel oil fired boilers to coal and nuclear fueled boilers and use of natural gas fired turbines to power electric generators. He is excited about the cost competitiveness and environmental benefits recently achieved by solar photovoltaic systems and the trends. Dan is a proponent of using solar energy in Arizona both at home and at work. After retiring from Arizona Department of Environmental Quality as an air quality specialist in 2007 he is using his experience at Fort McDowell's environmental department.

Contact Information for author:

Dan Catlin

Dan@catlinenergy.com

www.catlinenergy.com

phone 480-231-3703

SOLAR PHOTOVOLTAIC SYSTEM ECONOMIC ANALYSIS FOR THE ELECTRIC ENERGY CONSUMER

ALL OF THESE PRODUCTS ARE AVAILABLE THROUGH AMAZON.COM, CREATESPACE.COM, BARNS AND NOBLE, BOOKSTORES, AND BOOK DISTIRBUTORS.

WHOLESALE ACCOUNTS THROUGH SOUTHWESTPUBLICATION.COM, 3836 EAST DEWBERRY, MESA, AZ 85206 (480) 694-2871

Bigfoot in Arizona, Documentary, Part 1

Synopsis: On Memorial Day, 2008; MogollonMonster.com was formed to investigate Bigfoot sightings in Arizona. This Documentary video contains real research findings and may appear to be rough and fast is some spots. In some cases, you can tell the researchers are very nervous about the surroundings they have entered, Price: $11.95
Directed by Mitch Waite Runtime: 58 minutes Release year: 2011
Studio: MogollonMonster.com Productions   ASIN: B005BCRE2O (Rental) and B005BCRDRK (Purchase)

Bigfoot In Arizona, Documentary, Part 2 Close Encounters , Mitchell Waite (Director),

MogollonMonster.com (Studio) , List Price: $11.95, 60 minutes, NTSC, UPC: 886470301782 , The MogollonMonster.com Team takes on a major expedition to learn more about Arizona's Bigfoot. During a night ops event, they run into something that is not happy at them. Later in the night, something comes to their camp.

Bigfoot Shelters And Nests , Mitch Waite (Director), Mogollon Monster Studios (Studio) , List Price:

$11.95, 93 minutes, NTSC UPC: 886470337798 The MogollonMonster.com team has spent four years compiling data, photos, and video of Bigfoot homes, shelters and hunting blinds. Where do they sleep? What is their social structure like? And Much more.

Bigfoot Research 2009, Major Mitchell Waite (Director), MogollonMonster.com Productions (Studio), List Price: **$11.95**

93 minutes, NTSC, UPC: **886470563418**, Chronological documentation videos of Bigfoot research of the Mogollon Monster, Arizona's Bigfoot. These videos record the progress, findings, events and encounters of the MogollonMonster.com research team on the Mogollon Rim of Arizona.

Chasing the Mogollon Monster, Arizona's Bigfoot (DVD), Mitchell Waite (Producer), MogollonMonster.com (Studio), The Mogollon Monster Team (Actors), List Price: **$9.95**, 56 minutes, NTSC, UPC: **886470006137**, A selection of documentary film clips of Bigfoot researchers chasing the Mogollon Monster, Arizona's Bigfoot. Location is on the Mogollon Rim in Arizona. The team locates footprints, nests, scat, and experiences many vocalizations.

The Mogollon Monster, Arizona's Bigfoot, Authored by Susan Farnsworth , List Price: **$9.95** , 5" x 8" (12.7 x 20.32 cm), Black & White on White paper, 110 pages , ISBN 13: **978-1461016267** , ISBN-10: **1461016266**
BISAC: Nature / Wildlife, A collection of campfire stories of the Mogollon Monster, Arizona's Bigfoot as told by the Locals of Northern Arizona

More Mogollon Monster, Arizona's Bigfoot, Authored by Susan Farnsworth, Authored with Maj Mitchell Waite

List Price: **$9.95** 5" x 8" (12.7 x 20.32 cm), Black & White on White paper, 124 pages, ISBN-13: **978-1468064711**, ISBN-10: **1468064711** , BISAC: **Nature / Wildlife**, Part 1: More tales of the Mogollon Monster, Arizona's Bigfoot compiled from the locals of Northern Arizona. Part 2: Actual Bigfoot field research conducted by MogollonMonster.com

SOLAR PHOTOVOLTAIC SYSTEM ECONOMIC ANALYSIS FOR THE ELECTRIC ENERGY CONSUMER

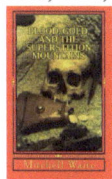
Have You Ever Seen A UFO?, Authored by Susan Farnsworth, List Price: $8.95, 5" x 8" (12.7 x 20.32 cm), Black & White on White paper, 100 pages, ISBN-13: 978-1461102397, ISBN-10: 1461102391
BISAC: Social Science / General, A selection of interviews with those who would know about UFOs--Our Military. Was Roswell real? Are there little green men?

Blood, Gold, And The Superstition Mountains, Authored by Mr. Mitchell Waite, List Price: $9.95, 5" x 8" (12.7 x 20.32 cm), Black & White on White paper 196 pages, ISBN-13: 978-1461096153 ISBN-10: 1461096154, BISAC: Fiction / Historical, An action/adventure thriller based on the legends and lore of the Lost Dutchman's Gold Mine and the Superstition Mountains of Arizona. Based on real people, places, events and treasure maps.

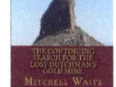
Blood, Gold, And The Superstition Mountains, The Return, Authored by Mitchell Waite List Price: $9.95 5" x 8" (12.7 x 20.32 cm), Black & White on White paper 134 pages, ISBN-13: 978-1461115502 ISBN-10: 1461115507, BISAC: Fiction / Mystery & Detective / General, An exciting action/thriller adventure story of the Lost Dutchman Gold Mine as it takes place in modern times. Based on history, real people, and events, this fictional novel will keep you glued to your seats till the end.

The Continuing Search for the Lost Dutchman's Gold Mine, Authored by Mitchell Waite List Price: $9.95, 5" x 8" (12.7 x 20.32 cm), Black & White on White paper, 152 pages, ISBN-13: 978-1461016229, ISBN-10: 1461016223, BISAC: History / United States / State & Local / Southwest, A study and research of the Lost Dutchman's Gold Mine in the Superstition Mountains of Arizona

Lost Dutchman Gold Mine Research And Related Stories, Authored by Maj Mitchell Waite List Price: $14.95 , 5" x 8" (12.7 x 20.32 cm) Black & White on White paper, 196 pages ISBN-13: 978-1466230385 ISBN-10: 146623038X, BISAC: History / United States / State & Local / Southwest, A unique twist on research for the Lost Dutchman's Gold Mine and the related stories of the Superstition Mountains

Desert Gold   Authored by Major Mitchell Waite, List Price: **$9.95   5" x 8"** (12.7 x 20.32 cm)   Black & White on White paper, 156 pages, ISBN-13: 978-1463777067, ISBN-10: 146377706X, BISAC: **Drama / American,** A tale of love, lost gold, and trechery in the Superstition Mountains of Arizona. A true western novel based on the legends and lore and history of Arizona.

Gold Panning Equipment, Build Your Own, Authored by Mitchell Waite   List Price: **$8.95   5" x 8"** (12.7 x 20.32 cm)
Black & White on White paper 80 pages ISBN-13: 978-1461135951, ISBN-10: 1461135958, BISAC: **Crafts & Hobbies / General,** Instructions and plans to build effective gold extraction equipment.

Backyard Gold Panning, The Perfect Part Time Job, Authored by Maj Mitchell Waite, List Price: **$16.95, 8" x 10"** (20.32 x 25.4 cm), Full Color on White paper, 74 pages, ISBN-13: 978-1470053512, ISBN-10: 1470053519, BISAC: **Sports & Recreation / Outdoor Skills,** Set up to gold pan in your backyard. Easy to do, and its a great part time job. With gold approaching $2000 per Troy Oz, it can be very profitable. This book tells how to pan, where to dig, various pieces of extraction equipment, and maps to go find your gold.

READING TREASURE MAP SIGNS AND SYMBOLS, Authored by Mitchell Waite Product Description, An in depth study of reading treasure maps symbol by symbol. The book also proposes solutions for several well known Spanish treasure maps and symbols found in the Superstition Mountains of Arizona. It goes even further and discusses cactus markers for treasure trails in the deserts of the Southwest US and Mexico. List price: $19.95 Paperback: 124 pages, Publisher: CreateSpace (July 1,2011, Language: English, ISBN-10: 1463685513, ISBN-13: 978-1463685515

ATVs, Build Your Own From Scratch, Authored by Mitchell Waite, Designed by Shannon Waite, List Price: **$12.95, 8.5" x 11"** (21.59 x 27.94 cm), Full Color on White paper, 46 pages ISBN-13: 978-1466485112, ISBN-10: 1466485116 BISAC: **Transportation / General,** Step by step instructions on how to build and assemble an ATV from scratch. Totally illustrated with diagrams and photographs. This ATV also features roll bars and cage for safety. Designed for a top speed of 45 MPH.

SOLAR PHOTOVOLTAIC SYSTEM ECONOMIC ANALYSIS FOR THE ELECTRIC ENERGY CONSUMER

The SKS 7.62X39 mm Rifle Disassembly And Cleaning Guide, Authored by Maj Mitchell Waite, List Price: $9.95

5.25" x 8" (13.335 x 20.32 cm), Full Color on White paper, 36 pages, ISBN-13: 978-1468119718, ISBN-10: 1468119710
BISAC: **Sports & Recreation / Shooting**, Complete disassembly and cleaning instructions for the SKS 7.62mm Rifle.

Esperanza Breathe, Authored by Norma Matheson , List Price: $9.95, 5" x 8" (12.7 x 20.32 cm),

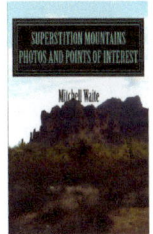

Black & White on White paper, 248 pages, ISBN-13: 978-1468091489, ISBN-10: 1468091484 , BISAC: Fiction / Christian / Romance
An action packed romance thriller written for Mormons and other Christian faiths. A story of a young woman's struggle with her drug biological drug cartel family she has never known. They learn of her existence and decide to eliminate her because of her new found FBI friend.

Superstition Mountains Photos And Points of Interest (NEW RELEASE!), Authored by Maj

Mitchell Waite, List Price: $14.95, 8" x 10" (20.32 x 25.4 cm), Full Color on White paper, 46 pages, ISBN-13: 978-1475214215, ISBN-10: 1475214219
BISAC: **Nature / Regional**, Full color photos and Topographic Maps of points of interest in the Superstition Mountains of Arizona.

The Round Valley Miracle, Authored by Maj Mitchell Waite, List Price: $8.95, 5" x 8" (12.7 x 20.32 cm), Black & White on White paper, 66 pages, ISBN-13: 978-1466270244 (CreateSpace-Assigned), ISBN-10: 1466270241, BISAC: Fiction / Religious, Cowboys and Angels. A story of the struggle the Pioneer settlers when called to settle the upper head waters of the Little Colorado River of the territory of Arizona in a pace called Round Valley.

BRAND NEW RELEASES!

Bigfoot Research 2010, Part 1

Mitchell Waite (Director), MogollonMonster.com (Studio) , List Price: $12.95 , 113 minutes, NTSC , UPC: 886470605095
Chronological video documentation of the Bigfoot research being conducted on the Mogollon Rim of Arizona. Video titles are:
Another Look at Bigfoot
Bigfoot Road Blocks
Bigfoot Game Camera Results
Mogollon Monster Hunting
Not a Bigfoot
Bigfoot Face Analysis
Bigfoot Nests and Guard Stations
Bigfoot Nests the Return
Bigfoot Hunting Techniques
Bigfoot Bed or Grave

Bigfoot Research 2010, Part 2

Mitchell Waite (Director), MogollonMonster.com Studio (Studio)
List Price: $12.95
113 minutes, NTSC
UPC: 886470608157
This DVD documentary records the events, findings, and theories of the MogollonMonster.com research team for the year 2010. The videos contained in this DVD are:
Bigfoot Campfire Discussion (8 videos)
First Sighting In June (2 videos)
Bigfoot Nest? Very Intriguing Footage
Bigfoot Nests, Old Root Cellar
Bigfoot Bed or Grave

Bigfoot Research 2010, Part 3

Waite (Director), MogollonMonster.com Sudio (Studio) , List Price: **$12.95**
113 minutes, NTSC , UPC: 886470610082 , Real Bigfoot Research conducted on the Mogollon Rim of Arizona. This DVD compiles the events, findings, and theories of the MogollonMonster.com research team. Videos within the DVD are:
Bigfoot Nest Very Intriguing Footage
Bigfoot Hunting Techniques
Mysterious Dirt Drawings
Bigfoot Equipment Checkout
Bigfoot Garden of Eden Project
Scouting New Bigfoot Area
Bigfoot or Bear
Bigfoot Road Hunting

SOLAR PHOTOVOLTAIC SYSTEM ECONOMIC ANALYSIS FOR THE ELECTRIC ENERGY CONSUMER

Bigfoot Sighting
Another Look at Bigfoot
Bigfoot Research, What Is It
Bigfoot Snow Tracking
Bigfoot Head Analysis
Bigfoot Research Camera Retrieval
Bigfoot Research Mounds

BRAND NEW RELEASES

Bigfoot Research 2010 Part 4

Mitchell Waite (Director), MogollonMonster.com (Studio), The Mogollon Monster Team Members (Actors) List Price: $10.95, 20 minutes, NTSC, UPC: 886470900954, 2010 season wrap up. The team obtains a not human fingerprint and collects some very interesting game camera footage.
Bigfoot Fingerprints'
First Bigfoot Outing 2010
analysis of Bigfoot Footage
Bushnel Game Camera Results
Searching For Bigfoot
Bigfoot, Wolves, and Coyotes, OH MY

Bigfoot Mischief
Bigfoot Research

Mr. Bigfoot's Photo Album

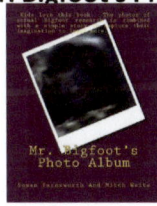

Authored by Susan Farnsworth, Authored by Susan Farnsworth, Authored by Mitch Waite
List Price: $9.95
8" x 10" (20.32 x 25.4 cm)
Full Color on White paper
34 pages
ISBN-13: **978-1478226826** (CreateSpace-Assigned)
ISBN-10: **147822682X**

BISAC: **Juvenile Fiction / Paranormal**
A collection of Bigfoot, and Bigfoot related photos in an album form. The photos are real (except one), and the script is mostly adopted for children. Adults and children both will enjoy the pictures and story line.

Lost Dutchman Gold Mine Research And Related Stories, Volume 2

Authored by Maj Mitchell Waite, Authored by Maj Mitchell Waite List
Price: **$35.95**
5.5" x 8.5" (13.97 x 21.59 cm)
Full Color on White paper
186 pages
ISBN-13: 978-1479153107
ISBN-10: 1479153109
BISAC: History / Expeditions & Discoveries
This book is the second of a six volume series covering the expeditions into the Superstition Mountains to search for the legendary Lost Dutchman's Gold Mine.

Lost Dutchman Gold Mine Research And Related Stories Volume 2 B&W edition:

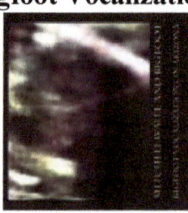

Black And White Edition Authored by Major Mitchell Waite, List Price: $14.95 5" x 8" (12.7 x 20.32 cm) Black & White on White paper 188 pages
ISBN-13: 978-1479189137
ISBN-10: 1479189138
BISAC: History / Expeditions & Discoveries
Volume two of six. This volume covers the research, findings, and expedition planning of a military and DoD civilian team searching for the Lost Dutchman's Gold Mine from June 1990 to May 1991.

Bigfoot Vocalizations in Arizona (CD)

List Price: **$9.95**
42minutes, 32 tracks
MogollonMonster.com
UPC: 886470908035
A collection of vocalizations believed to be those of the Mogollon Monster, Arizona's Bigfoot. It is believed by many Bigfoot can and does mimic animal calls such as birds, coyotes, and whistling, but the calls are slightly different from the real animals. Bigfoot also uses screams, moans, and howls distinctly their own.

SOLAR PHOTOVOLTAIC SYSTEM ECONOMIC ANALYSIS FOR THE ELECTRIC ENERGY CONSUMER

www.ingramcontent.com/pod-product-compliance
Lightning Source LLC
Chambersburg PA
CBHW041145180526
45159CB00002BB/737